REVIEW AND ANALYSIS OF:

Andrew Blum's Tubes: A Journey to the Center of the Internet

Review and Analysis Of Andrew Blum's

Tubes: A Journey to the Center of the Internet

REVIEW AND ANALYSIS BROUGHT TO YOU BY SUMMARY SHORTS

WWW.SUMMARYSHORTS.COM

COPYRIGHT PAGE

Copyright © 2016 by 'Summary Shorts' – Summary Media Inc

All rights reserved. No part of this book may be reproduced by any mechanical, photographic, or electrical process, or in the form of a recording. Nor may it be stored in a storage/retrieval system nor transmitted or otherwise be copied for private or public use-other than "fair use" as quotations in articles or reviews—without the prior written consent of the Author.

The Information in this book is solely for educational purposes and not for the treatment, diagnosis or prescription of any diseases. This text is not meant to provide financial or health advice of any sort. The Author and the publisher are in no way liable for any use or misuse of the material. No Guarantee of results are being made in this text.

Summary Shorts

Book title: Review and Analysis Of Andrew Blum's Tubes: A Journey to the Center of the Internet

—1st ed

Printed in the United States of America

Book Cover Design: 'Summary Shorts' – Summary Media Inc

Disclaimer

THIS BOOK IS AN UNOFFICAL REVIEW AND IS NEITHER APPROVED, ENDORSED, LICENSED, AUTHORIZED BY THE ORIGINAL PUBLISHER OR AUTHOR OF THE BOOK IN QUESTION. ANY INDIVIDUAL OR ORGANIZATION MENTIONED WITHIN THIS BOOK IS FOR REVIEW PURPOSES ONLY AND DOES NOT IMPLY SAID INDIVIDUALS OR ORGANIZATIONS ENDORSE ANY OF THE INFORMATION HEREIN. THIS BOOK ONLY SERVES AS A REVIEW AND IS NOT MEANT TO REPLACE THE ORIGINAL BOOK. IF YOU WISH TO READ THE ORIGINAL PLEASE GO TO AMAZON.COM.

Introduction

We use the internet for everything nowadays and surprisingly enough, hardly any of us know anything about the internet itself. We simply take it for granted. In Andrew Blum's book "Tubes" He goes into depth about the internet is and how deep it goes. We may only experience it through a browser, but it is much much more than that.

In this Summary, we will analyze, discuss and summaries the key points in "Tubes" By Andrew Blum. Enjoy!

Concept Summary

Concept # 1: The Internet Is a Real Physical Thing

Concept # 2: The Internet Truly Took Off In the 80s

Concept # 3: The Internet is made of millions upon millions of Networks

Concept # 4: Despite the millions of Networks, The Internet is controlled by a Small Group of People

Concept # 5: Most of the Internet is Underwater and In storage Centers

Concept Drilldown

Concept # 1: The Internet Is a Real Physical Thing

The internet is running all the time. Almost everything we use and do is influenced by it. Some people even have it running their home security systems. It has become more and more pervasive with time. Yet despite its enormous utility, hardly anyone knows anything about it. We simply take it for granted because it is always "on". So long as the light on the modem is on, we are good to go.

When we look at the router what we are seeing is the end point of a myriad of processes and connections. Our modem is connected by fibers which are connected to other fibers connected to huge data hubs who are in turn connected to other data hubs. You would think this would slow down the signal, but it actually helps it. The signal is made up of bits of light and its traveling faster than you can imagine.

It's quite hard to visualize just how this information is moving but if you want to see it in action, you can see it in real time at a data hub. These hubs are in nondescript buildings and rooms with thousands of blinking lights and cooling fans roaring 24/7 to keep all that data processing from overheating the servers. Some of the larger hubs are located in London, Tokyo, Frankfurt and Palo Alto. Despite these hubs, there are millions up millions of cables under the sea that connect the entire world as well. The internet is a lot more physical than it is virtual.

Concept # 2: The Internet Truly Took Off In the 80s

We tend to think of the internet as something that emerged in the mid to late 90s, but that would be incorrect, in fact, it started much earlier, in a primitive form that is. In 1969, the internet was online but it wasn't widely available. Only a few networks had it and those were mostly found in Universities. Universities used it to facilitate better communication between students and faculty. At that time only 5000 users had access and they were all documented in a "phonebook" if you will.

There was one problem with this earlier form of the internet though. Each network spoke a different programing language so the connections were not as seamless as they are today. That all changed in 1983 when something called TCP/IP became the standard for online networks. This universal online language allowed networks to communicate with relative ease. The invention of TCP/IP was a monumental event, because of it, the internet was starting to take off. By 1986 there were 400 networks online. Computers that had internet grew from 2000 in 1985 to 159,000 in 1989. Fast-forward and now it is so ubiquitous we do not even remember when we didn't have it.

Concept # 3: The Internet is made of millions upon millions of Networks

It's hard to believe that we now have 35,000 huge networks compared to just 30 years ago when we had hardly a handful of them. These giant networks are a thing of beauty. Despite their size, they are very well organized and the information flowing between them is virtually seamless. The bigger the network, the better it is. When you visit a website that information travels through those fibers we mentioned earlier. The more connections it passes through, the faster the information is delivered to you. This is where having many middle men works in your favor. The more middle men, the faster your information gets to you. These middle men are the hubs in the network. These hubs contain all the routers needed to facilitate the internet connections. The more routers and cables and tubes, the faster the internet is. Pretty neat.

Concept # 4: Despite the millions of Networks, The Internet is controlled by a Small Group of People

I am sure you noticed that some sites seem to load faster than others. In some cases the website itself has code that may slow it down or the server it is running on is slow for whatever reason. However, some sites run super-fast. Facebook for example will probably load faster than a local business site would. The reason for this is that the large networks tend to PEER with one another.

NANOG or North American Network Operators' Group essentially run the entire internet and port into one another's networks as to reduce the connection speeds. For example, Facebook has a very open Peer Policy often jokingly called "Peer slutting". They do this in good faith but they are not obligated to "peer" with another network hub. If a hub doesn't want you to connect, they can literally unplug your router from their hub. This will often cause internet slowdowns.

One of the most popular occurrences of this was when Sprint and Cogent pulled the plug from one another's hubs. As a result, the Department of Defense, The NY Court Systems and NASA couldn't send one another emails for nearly 3 days. Millions of normal

every day users of the internet also suffered.

Concept # 5: Most of the Internet is Underwater and In Storage Centers

Despite massive hubs connecting to one another's routers, the real internet load is running through cables found deep under the oceans. I guess in some ways you can consider these cables as the arteries of the internet. This isn't new to the internet, some 150 years, telegraph cables were also placed under the seas.

Very much like cables above ground, the ones in the seas can often get damaged due to natural disasters such as tsunamis or major earth quakes. One example of this was in 2006; A massive earth quake damaged a major section of the fiber cables in the Luzon Straits. As a result, most of China, Hong Kong and South Asia were knocked off line.

Much of the data we now store is in the cloud, which is made up of millions of servers and hard disks. Our data production is immense and it is for this reason they are so closely guarded in nondescript buildings. If they were to be destroyed, much of what we call the internet would be destroyed with it.

Summary

For better or for worse, we can simply not live without the internet now. Everything is wired and connected and there is little we can do about it. In saying that, it is always good to know what the internet is. It may not be practical to know all the nuts and bolts of what makes the internet tick; it's still good to know. Andrew Blum's book is one of the best books on the market that explains the internet in all its glory.

The Flow of the Book

Tubes, By Andrew Blum is a very good and incredibly easy read. It is written with all readers in mind. His explanations are well thought out and easily digestible. It's a wonderful and useful read.

Please see original book at http://amzn.to/28QrGRS

DID YOU ENJOY THE BOOK?

If you enjoyed this review, please let us know by leaving us a review. Reviews allow us to understand what we need to do to improve.

To stay tuned on future releases and to read free sample chapters, please visit our website at www.summaryshorts.com

ABOUT SUMMARY SHORTS

We at Summary Shorts understand that you are busy. We also know that you want to learn but simply don't have the time to read an entire book. We understand. It is our goal to give you the most concise review of books without sacrificing quality. Other review services tend to be light on information. We don't make that mistake. Summary Shorts Publications are available in several formats for your convenience. We are also going to offer other interesting and educational services on our site, including SHORT FACTS, a short video series teaching random facts from the books we review. We are also offering mini courses on various topics. Please check out our Website at

www.summaryshorts.com

ABOUT SUMMARY MEDIA INC

Summary media Inc, is a small publishing company specializing in History, Spirituality, Psychology and Health and countless other topics.

Founded by Doron Alon, a pioneer in the self-publishing industry, and a bestselling author of with well over 90 titles spanning several genres.

OTHER BOOKS AVAILABLE BY SUMMARY SHORTS

REVIEW AND ANALYSIS OF: BEN GREENFIELD'S 30 WAYS TO REBOOT YOUR BODY: A COMPLETE USER MANUAL FOR GETTING THE MOST OUT OF THE HUMAN BODY

REVIEW AND ANALYSIS OF: Charles H. Elliot's PhD and Laura L. Smith's PhD: Anger Management For Dummies

REVIEW AND ANALYSIS OF: Leonardo Lospennato's: The Da Vinci Curse: Life Design For People With Too Many Interests And Talents

REVIEW AND ANALYSIS OF: TREN GRIFFIN'S CHARLIE MUNGER: THE COMPLETE INVESTOR

Review and Analysis of Robert Greene's "Mastery"

Review and Analysis of: Alec Ross "The Industries of the Future"

Review and Analysis of: Gabriel Zucman "The Hidden Wealth of Nations: The Scourge of Tax Havens

Review and Analysis of: Drive: The Surprising Truth About What Motivates Us" Daniel Pink

Review and Analysis of: " Sonia Shah's Pandemic: Tracking Contagions, from Cholera to Ebola and Beyond"

Review and Analysis of: "Gary Taube's Why We Get Fat And What We Can Do About It"

Review and Analysis of" Andrew Blum's: Tubes: A Journey to the Center of the Internet

Review and Analysis of Bryan Caplan's The Myth of the Rational Voter: Why Democracies Choose Bad Policies

Review and Analysis of Eric Weiner's: The Geography of Genius A Search for the World's Most Creative Places from Ancient Athens to Silicon Valley

Review and Analysis of Mohamed A. El-Erian The Only Game in Town: Central Banks, Instability, and Avoiding the Next Collapse

Review and Analysis of: Aja Raden's: Stoned: Jewelry, Obsession, and How Desire Shapes the World

Review and Analysis of: Alan Watts's: The Wisdom of Insecurity: A Message for an Age of Anxiety

Review and Analysis of: Alex Cuadros's: Brazillionaires: Wealth, Power, Decadence, and Hope in an American Country

Review and Analysis of: Ayaan Hirsi Ali's: Heretic: Why Islam Needs A Reformation Now

Review and Analysis of: Benny Lewis's: Fluent in 3 Months: How Anyone at Any Age Can Learn To Speak Any Language From Anywhere in the World

Review and Analysis of: Carl Zimmer's: A Planet Of Viruses

Review and Analysis of: Dan Ariely's: Payoff: The Hidden Logic That Shapes Our Motivations

Review and Analysis of: Danielle LaPorte's: The Desire Map: A Guide to Creating Goals With Soul

Review and Analysis of: Marcus Aurelius's: Meditations

Review and Analysis of: Meltdown: A Free-Market Look at Why the Stock Market Collapsed, the Economy Tanked, and Government Bailouts Will Make Things Worse

Review and Analysis of: Pope Francis's: The Name of God Is Mercy

Review and Analysis of: The Bulletproof Diet: Lose up to a Pound a Day, Reclaim Energy and Focus, Upgrade Your Life

Review and Analysis of: The Ultimate Introduction to NLP: How to build a successful life

Review and Analysis of: Vishen Lakhiani's: The Code of the Extraordinary Mind: 10 Unconventional Laws to Redefine Your Life and Succeed On Your Own Terms

Review and Analysis of: Thomas M. Campbell II and T. Colin Campbell's: The China Study

FUTURE REVIEWS BY SUMMARY SHORTS

Review and And Analysis of: Cosmosapiens By John Hands

Review and And Analysis of: Freakonomics By Steven Levitt and Stephen Dubner

Review and And Analysis of: The Power of Habit By Charles Duhigg

Review and And Analysis of: Green Illusions By Ozzie Zehner

Review and And Analysis of: Emotional Intelligence By Daniel Goleman

Review and And Analysis of: Less Doing More Living By Ari Meisel

Review and And Analysis of: Micromotives and Macrobehaviors By Thomas Schelling

Review and And Analysis of: Six Thinking Hats by Edward de Bono

Review and And Analysis of: A Whole New Mind By Daniel Pink

Review and And Analysis of: Thinking Fast And Slow by Daniel Kahneman

Review and And Analysis of: Enchantment By Guy Kawasaki

Review and And Analysis of: Quiet By Susan Cain

Review and And Analysis of: The Wisdom of Psychopaths by Kevin Dutton

Review and And Analysis of: Predictably Irrational By Dan Ariely

Review and And Analysis of: Naked Economics by Charles Wheelan

Review and And Analysis of: Stumbling On Happiness By Daniel Gilbert

Review and And Analysis of: Mindset by Carol Dweck

Review and And Analysis of: Getting Things Done by David Allen

Review and And Analysis of: Flashboys by Michael Lewis

Review and And Analysis of: Crippled America by Donald Trump

Review and And Analysis of: Comfortably Unaware By Richard A. Oppenlander

Review and And Analysis of: Me, Myself And Us by Brian R. Little

Review and And Analysis of: Mindset by Carol Dweck

Review and And Analysis of: The Miracle Morning by Hal Elrod

And Many More. Please visit

WWW.SUMMARYSHORTS.COM to find out more.

www.ingramcontent.com/pod-product-compliance
Lightning Source LLC
Chambersburg PA
CBHW071223240526
45470CB00018B/2295